Published in 2021

Copyright © Stuart Owen, 2021

All rights reserved. No part of this publication may be reproduced or transmitted in any form or by any means, electronic or mechanical including photocopying, recording, or any information storage retrieval system without prior permission in writing from the publisher.

Photo credits

Sean Evett, John Faultless, Tam&Hazell Fergie, Mac Kac, Brian Look, Richard Lunt, Lee Maxey, Nathan Philpott, Barry Saunders Richard Tyrrell, David Wheeler, Iain Wilkins

Special thanks for their help goes to
John Branson and Michael Waters (Magpie Motorcycles and Scooters)

Introduction

By the late 1950s, and with a growing reputation, Innocenti had become a dominant force in scooter production. The Lambretta LD boasted reliability, comfort, and style, but unfortunately was becoming a little outdated. Its successor would need to be something different, not only in terms of performance but also in looks. After several years of development and testing a new series was ready to take its place in the shape of the Li and TV.

Both models were totally different from anything Lambretta had built before offering new and innovative ideas that took the market by storm. With a more powerful engine, that was far easier to access and maintain, it gave the Lambretta the performance required to cope in modern-day traffic. Complimented by a revised chassis layout and improved braking system handling was greatly improved. The new shape and bodywork styling meant there was no other scooter like it making both models an instant hit with the buying public. So good was the design, of the Li and TV, that all future Lambretta models would follow a similar layout right up until the end of production in 1971.

History

The Lambretta LD was in its third mark by 1957 and though it had served well during its six-year tenure was now becoming a bit outdated, both in its engineering and design. A replacement was well on the way having started development two years before and was ready to go into production in the autumn of that year. It had been a decade since the first Lambretta had rolled off the production line and each successive model had followed along the same lines. A shaft-driven three-speed engine that was fixed to the frame with a torsion bar suspension, 8-inch wheels with small drum brakes, and basic bodywork, even though it had been improved significantly as the LD progressed.

To take the Lambretta forward and meet the demands required to cope with modern-day traffic meant a radical rethink of the design. The central component of any two-wheeled vehicle is the engine and that is where wholesale changes would be made. To start with the cylinder would now be horizontally positioned compared to the traditional vertical layout. This was a new idea, but in doing so opened a huge space above the engine which could house a larger fuel tank and air intake system. The transmission, which up to that point had been shaft-driven, would now switch to a chain and with bigger casing dimensions house for the first time a four-speed gearbox.

An early press shot of the new TV 175 which hailed a new era in Lambretta design

The one downside of the horizontal engine design was that the exhaust port would face towards the ground, not leaving much space for the exhaust system to fit in. In fact, with the current 8-inch wheel configuration there wouldn't be enough room to house it, and certainly not without it grounding on the surface of the road. The answer was simple enough, and so the wheels would now increase to a 10-inch diameter. There were other benefits of this change most notably with the hubs which with their bigger size allowed for vastly improved brake surface area. The slightly longer wheelbase would also give a greater improvement in handling, which was important as the engine would be producing significantly more power.

The final part of the engine that differed was in the way it was mounted, the new design meant that it would be centrally pivoted between the cylinder and the transmission. There would be one offset suspension unit fixed above the gearbox and mounted at a 45degree angle to the top of the frame thus allowing the front of the engine to face slightly upwards giving more clearance for the exhaust system. The main advantage was a much lower centre of gravity and combined with better suspension damping at the rear resulted in a much smoother ride and greatly improved handling during cornering.

The new chassis layout was longer and taller, but the width was slightly narrower, giving the new shape a much sleeker look about it compared to the LD. The top shell did away with the step at the back and was replaced with a gentle curve to the rear light unit, this allowed for much bigger side panels and transformed the back end of the machine. Underneath the panels, a central frame tube lay horizontal above the engine and left a space to house a much larger fuel tank and toolbox, with the airbox arriving later in production. The new layout was a huge improvement on previous models and would benefit the customer no end.

The bodywork styling at the front used the traditional leg shield design, but the front mudguard was for the first time fixed compared to a turning one on all previous models. One of the reasons was down to the new front fork design, which was much bigger and robust and now housed a 10-inch wheel. Another noticeable change was the headset which was completely enclosed offering better weather protection for the controls, even though the headlamp would remain fixed in the centre of the horn casting like that of the LD. The result of all these changes gave the Lambretta a completely new look and feel about it and lined up against the LD it looked much bigger and more powerful, which it was. Innocenti had invested heavily in the new design as it required new tooling and production lines at the factory, so they needed strong sales from day one. It was now down to how the public accepted it to determine if it would be a financial success.

The first or the new generation Lambretta's to go on sale was the TV 175 which would line up in showrooms against the LD. Though it was significantly more expensive originally being offered at £209.00 it was a far superior machine. Both Innocenti and Lambretta Concessionaires were confident it would appeal to the public despite its hefty price tag because of the performance it offered. The real target was existing Lambretta owners who would be keen to upgrade to what was at the time the most powerful Lambretta available.

The first advert in the UK was rather sedate just announcing the TV 175 had landed. That soon changed with slogans emphasising the machines performance

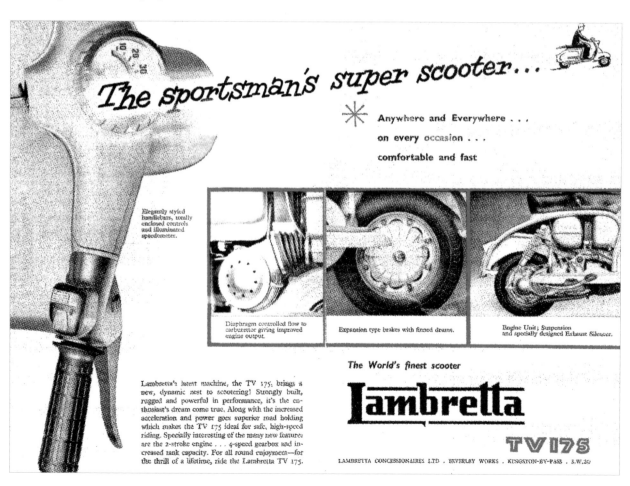

It was soon targeted as a sports scooter which in essence it was easily outperforming anything in its class

There were three new models to choose from, the Li 125, Li 150, and the TV 175. Both Li models would use the same engineering throughout, but with different power output, the 150 version being slightly faster. The TV 175 however, would have its own unique engineering both from a bodywork and engine perspective, the engine design being completely different and a far more complicated setup compared to the Li. For now, that didn't matter, but as time went by it would expose a flaw in Innocenti's thinking and at a huge financial cost. The first machine to go on sale was the TV 175 in September 1957, and as usual, launched to the customary big fanfare surrounding anything new in the motorcycle and scooter industry. UK customers would have to wait slightly longer, with the first ones arriving in showrooms at the beginning of February 1958 and a further wait of up to six weeks for delivery after purchase. The sale price would be £209.00 which compared to the LD was a significant price rise, remembering the LD was still available.

Publicity shots were based on the superiority of the TV 175 and quite often an LD which it was replacing would be shown in the background suggesting its time was up

The difference in performance was huge and the 8.6bhp engine could give a claimed top speed of 64mph, compared to the LD with 6.0 bhp and 50mph top speed the gulf was immense. The improved braking and suspension, including front dampers on the forks, meant anyone looking at it from a technical point couldn't fail but to be impressed, and with its new styling and a dual seat those that wanted the looks would also be enticed. It would be made available in just one colour, ivory, but combined with the grey trim and floor mats offset with a black finish on the seat they complimented each other perfectly. As per normal Lambretta Concessionaires went into overdrive advertising it, showcasing its good points in their sales drive. "The sportsman's super scooter" and "powerful and speedy" were just two of the slogans used which compared to what else was on offer was entirely true. There was a worry the sales of the LD would nosedive as customers flocked to the new Lambretta and that was confirmed when its price was reduced to just £130.00. With the impending introduction of the Li, it would make it even harder for dealers to shift them, so the rush was on to clear remaining stock from showrooms as quickly as possible.

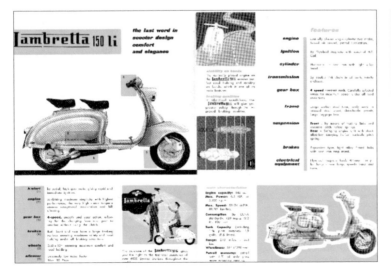

Compared to the TV 175 the Li on the other hand was rather more sedate in its approach

Rightly so, as in April the Li 150 was announced to the public with the Li 125 following in June. Both machines were already being heavily publicised and with huge interest, many owners of older models were ready to upgrade as soon as they were available. The 125 version was priced at £155.00 and the 150 for £174.00, making them hugely competitive against their rivals. The 125 with its modest 5.2bhp and 50mph top speed may have been similar to that of the LD, but with its four-speed gearbox being far more tractable especially uphill. The 150 had better performance figures of 6.5bhp and 53 mph top speed combined with greater acceleration, and it would be this model that was the most successful in terms of sales, a trend that continued when later series were introduced.

The UK press release for the Li 150 was signed by Peter Agg but didn't give too much away in relation to technical and performance specification

The UK press launch was conducted By James Agg Snr and featured distinctive cutaway sections on both the front and rear hub

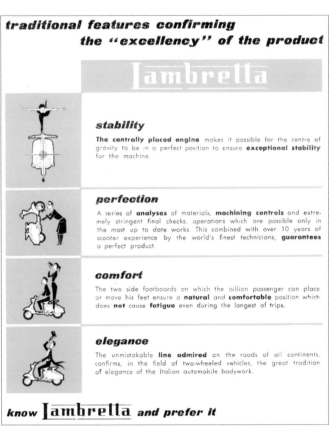

The new chassis design and bodywork styling were highlighted on all literature to get the point across just how good the improvements were

The advertising budget to promote the Lambretta was still growing and Lambretta Concessionaires realised they had a winner on their hands so raised the coverage even further. It wasn't just existing customers, but new ones that were fuelling the market as the two-wheeled industry continued to grow at a huge rate in the UK. The introduction of the Li and TV couldn't have come at a better time and sales had the potential to skyrocket if marketed correctly. To gain maximum exposure Peter Agg came up with an idea that took the whole industry by storm. Instead of going direct to the national press trying to get coverage he turned the tables on them and invited them to come to him instead. Taking approximately 150 machines, a mix of all three models, anyone from the motoring and national press was invited down to Crystal Palace racetrack to take one for a free test ride. Assembling the company's top sales and technical staff to be on hand to ask questions it was a bold marketing attempt to take on and one that paid off as the reports came flocking in one after the other giving the Lambretta huge media exposure at no extra cost.

Members of the motoring press putting all three new Lambretta models to the test at Crystal Palace racetrack

The Li 150 sold itself with ease whereas the Li 125, the basic model had press shots glammed up to make it more appealing

With all three models offering such good all-around performance the reports were all favourable and a glowing testament to their design. This resulted in a huge increase in enquiries which translated into unprecedented sales, so much so that the company could hardly keep up with demand. This had a knock-on effect in Italy as Innocenti had to ramp up production to keep the supplies flowing without interruption to the UK. The worst-case scenario would be a lack of availability in showrooms, but thankfully both companies worked hard to make sure that never happened. While the Li models were a resounding success the TV 175 did struggle in comparison, so its price was reduced to £199.00. There were also other issues surrounding it that were a concern mainly to dealers. Part of the agreement to be an official dealer required them to carry a huge stock of spares, both for repairs and aftersales, this was standard procedure across the industry, but the problem was the unique engineering of the TV 175. Both Li models used practically the same components throughout so only required one set of stock between them. The TV 175 needed a separate stock of its own, thus incurring extra cost to the dealer and the problem many feared was what would happen when the model was discontinued, would its successor use similar engineering?

Another problem was the cost of the tools required for servicing it, dealers had to carry separate tooling, and not forgetting staff required different training to be able to work on them. Compare that to both Li models that used one set of tools between them and didn't require extra training as the layout was identical. For some time, dealers had been expressing their thoughts on this to Lambretta Concessionaires, but it was Innocenti who had come up with the design, so they dismissed the claims. The engine on the TV 175 was quite experimental in many ways and some of the components had a complicated setup in their making. It was a reliable engine but required a competent level of mechanical skill when it came to working on it. Innocenti never explained why it was designed this way and that everything on it, not just the engine, wasn't compatible with the other models. The LD, and subsequent models before it followed along similar lines when it came to the engine, but this was different and wasn't a revamp of something existing so was going to be unique. The Li had had a similar scenario but the engine was far less complicated and much easier to work on not forgetting cheaper to produce. With fewer components, its reliability was going to be even better, in fact when introduced the Li engine was a revelation due to its simplicity. To produce a two-stroke engine that was chain-driven with a four-speed gearbox and be so compact was a design master class. So, the question begs if Innocenti had already produced this engine why did it need to go down the route it did with the TV 175? Surely it would have been far easier to use the same design just with a bigger capacity.

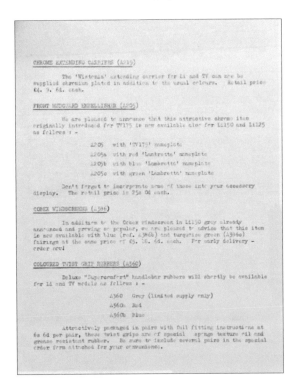 

Lambretta Concessionaires offered a wide range of accessories for both the Li and TV but there was outside competition. This was heavily frowned upon, and senior staff warned dealers found stocking aftermarket equipment from other companies that they could have their dealership taken away from them

Sales of all three models had varying levels of success and proved to Innocenti that in the future things would need to be done differently. The Li 125 sold over 47,000 units which made it successful but paled into insignificance compared to the Li 150 with official figures stating 110,000 being sold, more than double that of the Li 125. It was the TV 175 that was a problem with a total of just 10,000 being built, not even a tenth of the Li 150, so it was no surprise that in December 1958 the last one rolled off the production line. Both Li models would continue for almost another year, with the last ones being produced in October 1959, so successful was the Li range that it would continue with the Li series 11 replacing it the same month. That is when the Li became known as the first series, once its successor was named. Though the styling of the bodywork would alter the tooling remained the same, similarly with the mechanical components. There were developments to the engine in the way of upgrades, but everything was interchangeable.

By 1959 the Li was a huge success and led the way for the Lambretta in the UK. Though the TV 175 was still being offered by this time its days were numbered. Due to its different engineering, Innocenti realised they needed to bring it in line with the LI models so its replacement the TV 175 series 11 would do just that

 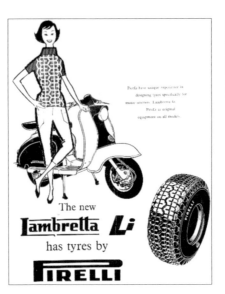 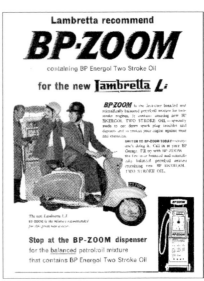

Many of the companies whose brands backed the Lambretta had chosen the Li over the TV when it came to promoting their products

Where did that leave the TV 175? Production of it had ceased much earlier, but it too had a successor in the TV 175 series 11, even so, it was far removed from the series 1 version. Its engineering came in line with the Li range, not just the mechanical side but also the bodywork, that way all three models would be compatible with each other and make it far easier to manage them. It also brought home the truth that many dealers feared, as the stock for the TV 175 wouldn't be compatible with anything else. This meant that they were sitting on a huge number of spares that became almost worthless overnight. The only saving grace was this wouldn't happen in the future, a claim made to dealers by Lambretta Concessionaires to gain their trust back and one that was adhered to as all existing series remained compatible throughout their model range right up to the Grand Prix and the end of Lambretta production in 1971.

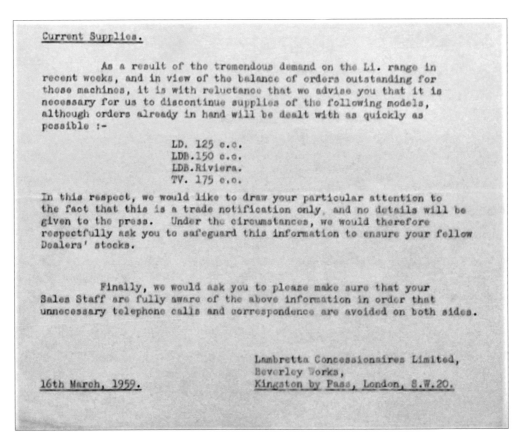

A press release from Lambretta Concessionaires dated 16th March 1959 stating that all models except the Li range would be discontinued bringing down the curtains on the TV 175

Lambretta Newsletter

SALES & ADMINISTRATION

To: All Lambretta Dealers   No. 62/59

Li. 150 Series I Model.

You will no doubt recall our Newsletter No. 49/59 of the 24th October, wherein we introduced the series II Li range of Lambretta Scooters, at which time, due to the general demand for machines, it was felt to be advisable to continue both the series I and the series II models side by side. This was done to minimise the delivery delays, and at the same time offer to the public the optional positioning of the headlamp assembly.

From information received it would appear that the demand for the series I Li.150 has decreased and, in fact, many Dealers with stocks of these machines are finding them to be an embarrassment when shown alongside the series II type.

We are therefore stopping production on the <u>series I Li.150 model</u> and, in order to assist Dealers we propose to supply for each Li.150 series I machine invoiced by us on or after the 1st September 1959, and still remaining on Dealers' premises brand new and unsold, the following items, FREE OF CHARGE -

    A. 212 - Rear carrier.

    A. 386 - Cobex Windscreen.

With these two popular accessories fitted, we feel certain that Dealers will then find no difficulty in selling these machines to the public.

To obtain the two accessory items, mentioned above, the following information must reach this office within the <u>next 7 days</u> -

    <u>Engine No.</u>   <u>Frame No.</u>   <u>Colour.</u>   <u>Invoice No. & Date.</u>

On receipt of these details, our Accessory Department will immediately put in hand the delivery of the carriers and windscreens, but we would stress that this offer applies to the <u>Li.150</u> series I machine only. It is not felt necessary to include the Li.125 series I model due to the considerable difference in price between the series I and the series II types.

We feel sure you will welcome the above gesture, and should any Dealers require further stocks of the Li.150 series I, supplied with the accessories mentioned, please let us know by return as stocks of these are limited.

                                    D. G. Miller - Sales Manager.
                                    Lambretta Concessionaires Limited,
15th December, 1959   Purley Way,
                                    Croydon.

The end of the Li range with the announcement in December 1959 Lambretta Concessionaires would stop production. To help move remaining stocks a free windscreen and rear carrier were being offered to entice customers

In Italy sales leaflets were now offering both Li models and the new TV 175 series 11 at the same time

Riders from New Zealand and Australia on tour in the UK at the Lambretta Concessionaires factory showing what a worldwide success the LI and TV had made the Lambretta

An Li 150 being put through its paces at the Isle of Man rally in 1959, further evidence of just how good a machine it was. Proving not only reliable but competitive against other makes of motor scooters

The TV 175 was a ground-breaking machine and, though it had its faults, was the model that moved Innocenti and the Lambretta forward. Without it, it might never have happened and so rightly deserves its place in Lambretta history. Lessons were learned from it and thus gave the company a clear path to follow after its demise. For the Li range, it proved to be the real winner with its uncomplicated design, but great looks and reliability. The series that followed became even more successful as it became the backbone of Lambretta design for many years. The li 125 and 150 were pioneering machines in many ways and should be regarded as one of the finest designs ever in the history of two-wheeled transport.

The Scottish Six Days Trial

The Scottish Six Days Trial is held annually and is a gruelling endurance event that takes riders across some of the most rugged terrain of the Northern highlands. Taking place over the first week in May and covering several hundred miles many riders and their machines fail to make the whole journey, so it came as a surprise that in 1959 Lambretta Concessionaires decided to enter the competition using three modified Lambretta Li 150 machines. It was seen by the industry as a cheap PR stunt that was destined to failure, but those working on the machines at the factory knew otherwise, as they were serious and dedicated about the task ahead spending many months carefully planning and preparing for the event.

Of the three riders who entered Alan Kimber was the one who went on to greater glory. Over the next few years, he entered various trials on a Lambretta and was chief test rider of the Rallymaster prototype playing a significant role in its development. In a strange twist, Alan worked at Lambretta Concessionaires before leaving to take on the role of running Suzuki GB in the mid-1960s. With sales poor, it was wound up and Peter Agg, Alan's former boss at the Croydon factory took control of the stricken company making it a huge success in the 1970s

Each machine had its front bodywork stripped down with the leg shields and front mudguard replaced with those from a Model D. To help absorb some of the shocks from the uneven surface a pair of front dampers were fitted to the forks, similar to those on a TV 175. At the rear, the splash guard was removed and a specially modified exhaust was made that went up and over the engine to give as much ground clearance as possible. All three were frame breather models which helped as the clean air was sucked in from the back of the frame as opposed to the front which could get obstructed by dirt and debris from the off-road track sections of the course. A spare wheel was fitted at the rear of the frame and a quick fill gas bottle installed by the rear mudguard, crudely strapped to the left-hand rear footboard strut. Despite the rugged terrain that would be encountered all three machines were only fitted with road-going tyres making the challenge even more difficult.

The entry list for the trials with the three Lambretta riders down as numbers 3,7 and 8. Though the attempt was done by Lambretta Concessionaires they are listed under J.R. Alexander & Co.Ltd Scotlands biggest Lambretta agent at the time. Despite the huge success of all three riders competing the event, Lambretta Concessionaires didn't make too much noise about it with just this single advert announcing their success

The three riders selected were Alan Kimber, Lewis "Ludo" Moore, and Geoff Parker. Upon entering onlookers and fellow competitors didn't give them any chance of succeeding with most presuming they wouldn't even last the first day. Gordon Jackson who was a seasoned pro at the event, and that year's winner, was invited to inspect the machines and wished them all the best, but knowing the track so well doubted they would make it to the end. Despite the tough challenges and almost impossible conditions all three completed the six days of the event and though repairs were needed after each stage there were no breakdowns reported. Lambretta Concessionaires had proved both the Lambretta's reliability and strong build qualities using it to maximum effect to gain media exposure. Keen to build on this success the event was entered again in 1960 and 61 as well as other off-road trials, with Alan Kimber remaining as one of the riders, but this time using a TV 175. He along with other mechanics at Lambretta Concessionaires used these events to develop and test the Rallymaster which was introduced during series 11 production and based on a modified Li 150.

The continuation of the Li and TV story

The Lambretta has many enthusiasts devoted to keeping the name going by way of preservation and the Li and TV 175 are no exception with many dedicated followers who have produced some stunning restorations.

This immaculately preserved Li 150 and Bambini combination is a great example of how important the accessory market was. With the shape different to the LD new designs were required and it became a very lucrative market both for Lambretta Concessionaires and outside companies who they frowned upon as it was seen as taking their trade away

Though it's hard to find them with paintwork still in its original condition by leaving it as it is with a few little additions the great design and character of the Li shows through

By adding a different seat and small sprint rack on the back of the frame it adds an interesting take on the Li. It also makes you wonder if Innocenti could have tweaked the design slightly what could have been

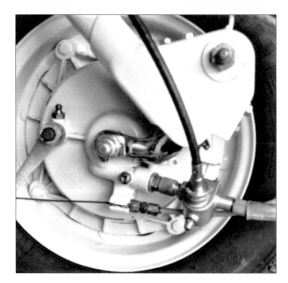

This Li 150 is a tribute to the late Geoff Parker who competed in the Scottish Six Days Trial and several other events for Lambretta Concessionaires in the late 1950s. Featuring replica modifications of how they were originally done, it's a stunning recreation of one of the very few competition machines based on the Li

The famous Wembley 9's Li first featured in the Richard Barnes book "Mods" Thought lost forever it was rediscovered in a London basement and returned to its former glory

With its extra power compared to the LD the Li was the perfect choice for either a sidecar or trailer. Though not specifically made for the Lambretta the PAV trailer is the ideal accompaniment when it comes to stowing away extra luggage if touring and looks well suited to the Li. The TV 175 was only painted in one colour by Innocenti, but had it been available as a two-tone option it could have been much more striking as this restored one clearly shows

The TV 175 with its unique features is an ideal base for individual customising, but without taking its original lines away in this example. The addition of original leopard skin style accessories only further enhances its looks

Painted in the coffee and cream combination offered by Lambretta Concessionaires this restoration sports many aftermarket accessories including working indicators. It was this style of making additions that were the roots of early Lambretta customisation as owners wanted their machines to look different from everyone else

Production specifications and changes

Though all three models were only in production for a short while, the TV 175 fifteen months, and the Li 125 and 150 sixteen months they did go through several production changes. These were mainly mechanical with some slight cosmetic alterations also.

Li 125-150

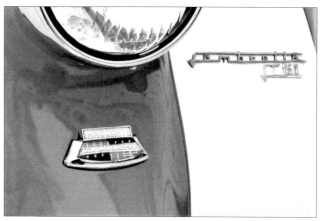

The Li 125 had a 50mph speedometer fitted and the 150, 60mph (pictured). The Lambretta badge script at the front was identical the only difference being the Li badge which was either 125 or 150

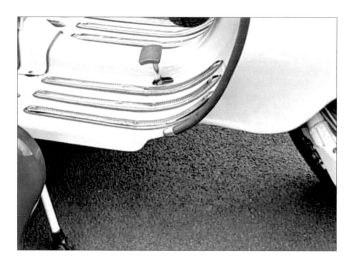

The 125 had one-piece aluminium floor channels fitted while the 150 used the traditional rubber inserts and endcaps

Both models used a single non-adjustable chain tensioner which was changed to the two-piece adjustable type later in production

The original frame breather system on both models changed over in late production to the frame top airbox system that was used throughout the rest of Lambretta production

The air was sucked in through the rear grill at the back of the frame. Despite early examples having an air vent cut out at the bottom of the side panels by the time they had arrived in the UK, this had been done away with. The picture on the right showing the press launch machine without the vents

The original Innocenti X-ray factory drawing shows how the rubber cover sucked air in from the rear of the frame. It also shows the compact chain-driven transmission and horizontal engine layout that was a revelation when it first appeared to the automotive press

The Li 125 came in just one colour of light grey (Grigio 8019) seen here in this factory promotional image. Later in production, it had dark grey side panels (Grigio 8040)

The Li 150 had greater options when it came to paint colours with the frame in grey (Grigio 8019) and the side panels horn casting and headset top offered in the following Azzurro 8032, Bleu 8031, Verde 8015, Rosso 8046, and Rosso 8047. Note: The Li 150 on the right is regarded as a crossover model where the intake is now through the top of the frame into the airbox but still has the open vents at the back of the frame

TO All Lambretta Dealers.   No. 2/59.

Reduced Gear Li.150 Model.

          Over the past few months we have received many requests from both the trade and the public for an Li model Lambretta adapted for sidecar work, and we now have pleasure in advising you that due to the design of the machine, the simple operation of withdrawing the gear cluster from the Li.150 model, and replacing it with the Li.125 gear cluster, has the effect of converting the 150 c.c. machine to the ideal ratio. As Dealers will appreciate, this conversion is relatively simple, and could easily be carried out in their own workshops.

          We would also draw your attention to the fact that it is possible to order a new Li.150 machine ready-fitted with reduced gears, at the same price as the normal Li.150 model, but we would stress that orders must be clearly marked "reduced gear model". At the present time the following colours are available from stock :-

          Grey/red; Grey/blue; Grey/turquoise; All Red.

A UK factory press release from early 1959 for fitting the Li 150 with an Li 125 gearbox as it had a better 1st gear ratio for pulling away when a sidecar was added. Also mentioned are the colour combinations that were available at the time

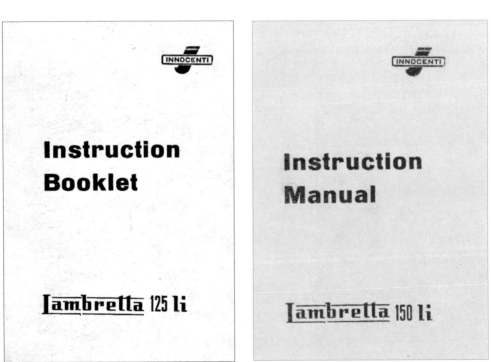

The li 125 instruction booklet came in a light grey colour like the look of its colour scheme. The Li 150 version was far brighter in its blue colour

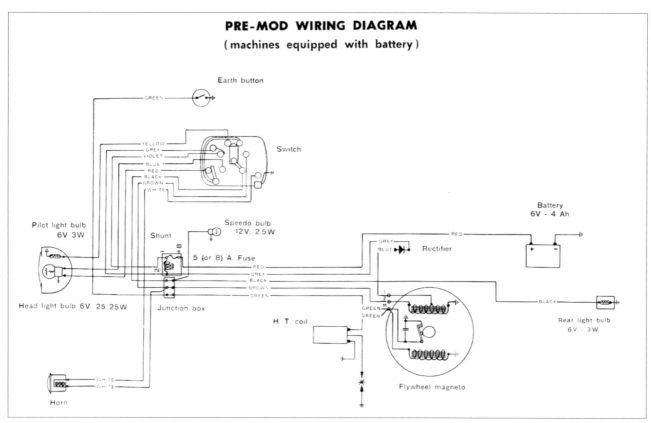

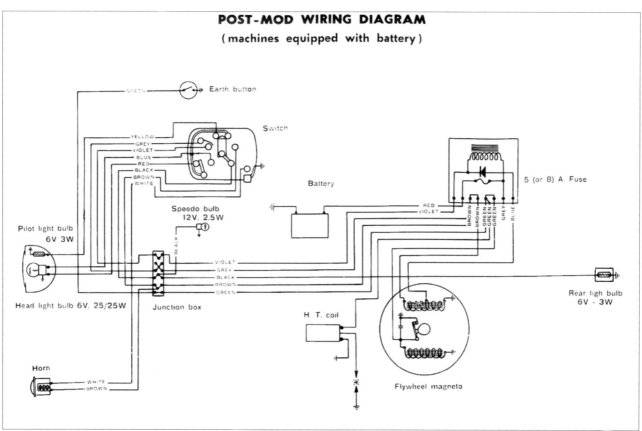

Li 125 and 150 pre and post-mod wiring diagrams. Pre modification has the resistor and fuse fitted behind the headset junction box whereas the post-modification has them fitted with the rectifier and positioned on the left side footboard strut

TV 175

The TV 175 was the first Lambretta to feature a 70mph speedometer and its bezel had a serrated edge unlike the plain one on the Li. The Lambretta script badge was similar to the Li, but the TV badge was done in capitals

 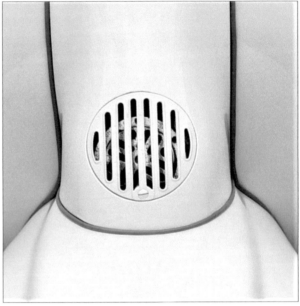

The first version of the horn casting was a one-piece design. The second version had a removable grill making access to the horn much easier

Two views of the engine-transmission show how much the design differed from the Li

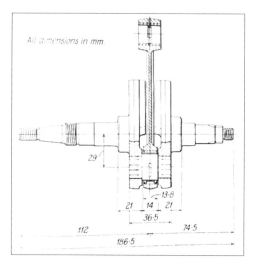 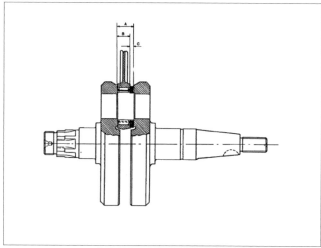

On the left the LD crankshaft which was much smaller compared to the robust dimensions of the TV 175 version. Innocenti realised with the extra power the new engine created it was vital to make it strong enough to cope

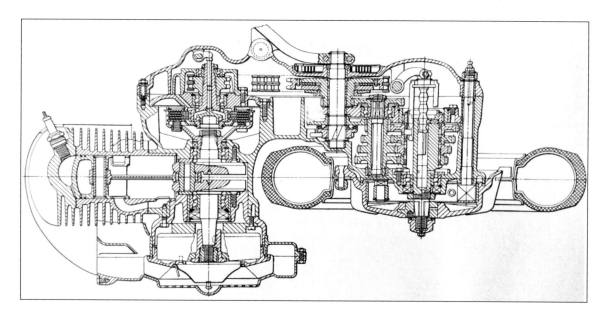

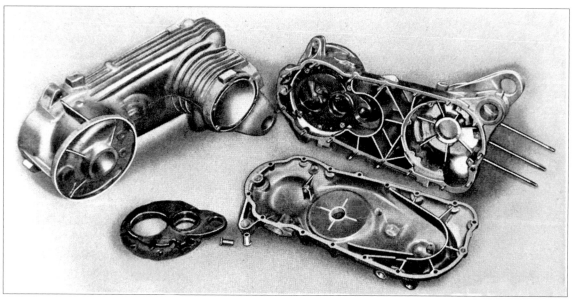

The TV 175 engine was compact but nowhere near as much as that of the Li and was far more complicated in its design

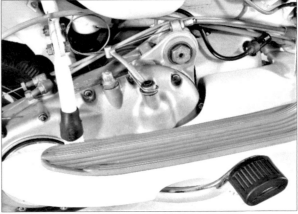

The induction was the same as on the Li through the frame but used the bigger Dellorto MB23 BS5 carburettor. The clutch was positioned at the front of the transmission and the kickstart centrally

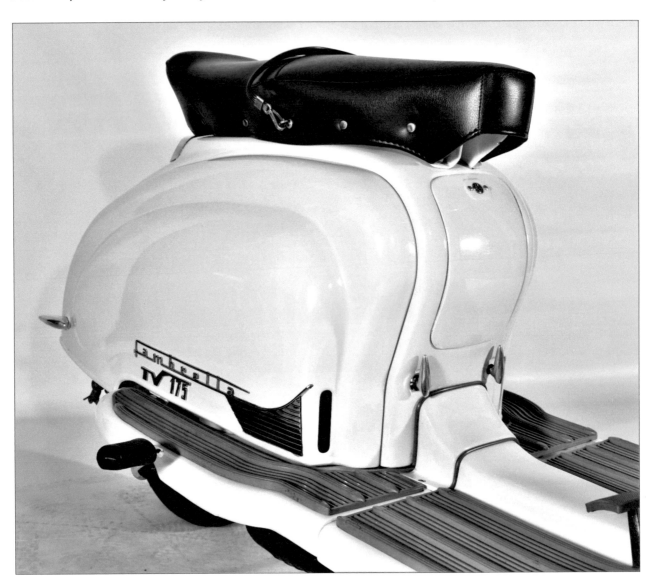

The unique panels of the TV 175 had air vents at the front throughout its entire production, unlike the Li which only had them early on. Because the kickstart was centrally positioned there was no need for a cut-out on the panel for the kickstart pedal. Note the twin-seat the first time a Lambretta had been fitted with one. Instead of floor channels, thick rubber floor mats were fitted instead both on the leg shields and footboards

From the front, the new sleeker lines were aided by the fixed front mudguard. At the back the taller frame shell made it stand out so much better than the LD, the vents for air induction are clear to see

It might have been the biggest and heaviest Lambretta to date, but was also the most powerful by a significant margin

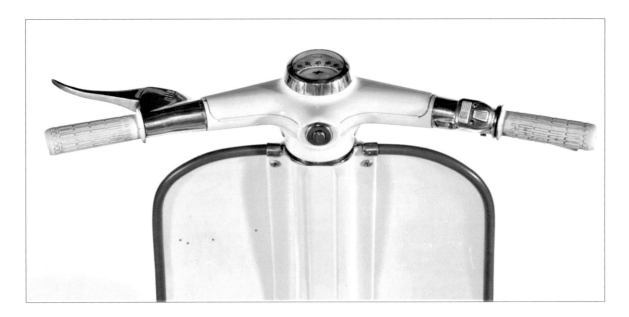

The TV 175 was the first Lambretta to incorporate a fully enclosed headset, Note the central ignition switch with the searing lock on the opposing side

Cable tension on the clutch and the front brake was done by way of an adjustment on each side of the headset. This was phased out in later production. The front brake was the first Lambretta to feature external dampers. Note the chrome finish on the outer of the wheel rims

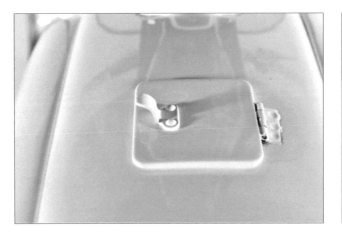 

The petrol flap featured a handle to lift it instead of a raised edge. The dual seat was made by Italian company Aquila Continentale

Innocenti only produced the TV 175 in one colour, ivory (8028). However, Lambretta Concessionaires were more ambitious by offering it in a two-tone scheme of Ivory and coffee. There were only ever two press releases referring to this extra choice the one below referring to paint spray for dealers

PAINT SPRAY TV IVORY/CREAM (A94h)

We are pleased to advise that Ivory/cream paint sprays at 10s 6d each retail are now in stock. This brings the range of paint sprays to : -

Red (A94a) suitable for LDB, LDA and pre April Li150

Winchester blue (A94b) for LDB and Li150

Mk III off-white (A94f) for Mk III LD125 and LDB

Slate blue (A94g) for Mk III LDB and LDA

TV Ivory/cream (A94h) for TV175 Series I and Coffee/cream

Li150 grey (A94r) for Li150

TOUCH-UP PAINT

Please note that new Lambretta colours have been added to the range as follows : -

A91u	⅛ pt.	Flaminia grey	2s 0d each retail
A91v	⅛ pt.	Glacier Blue (Azure)	2s 0d each retail
A91w	⅛ pt.	TV Yellow	2s 0d each retail

```
TV.175 Model  -  IMPORTANT NOTICE.

              As a result of numerous enquiries from our Dealers
for additional colour schemes for the very popular TV.175 model, we are
pleased to announce two new dual-colour versions, as follows :-

              Ivory / Ocean Green.
              Ivory / Winchester Blue.

As supplies of these new colours will be rather limited in the first instance
orders will be dealt with in strict rotation.  We would also make it clear
that the already well-known Ivory and Coffee/cream versions of this model
will still be readily available.


Li. Sectional Drawing.

              We are enclosing with this Newsletter, a sectional
drawing of the Li.150 model Lambretta, which we know you will find of
interest, and would suggest that, suitably mounted, this would be an ideal
display item for your showroom or reception area.



                                        Lambretta Concessionaires Limited.
                                        Beverley Works,
18th February, 1959.                    Kingston by Pass, London, S.W.20.
```

As well as the ivory and coffee option Lambretta Concessionaires offered dual colours of ivory with either ocean green or Winchester blue. This press release was dated 18th February 1959 and after production had finished in Italy. It is presumed by this time there were limited stocks left in the UK, so it is not known how many were sprayed with these options

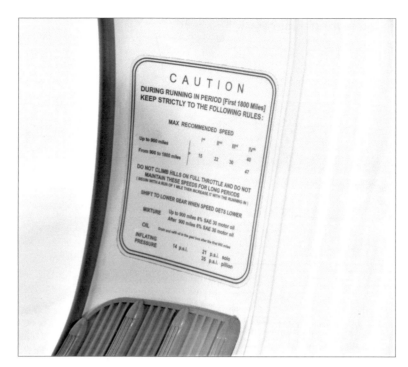 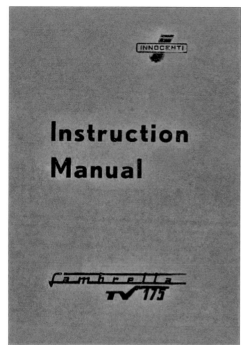

Running in instructions were placed just above the floor rubbers on the left-hand side of the leg shields. The instruction manual came in a burnt orange front cover

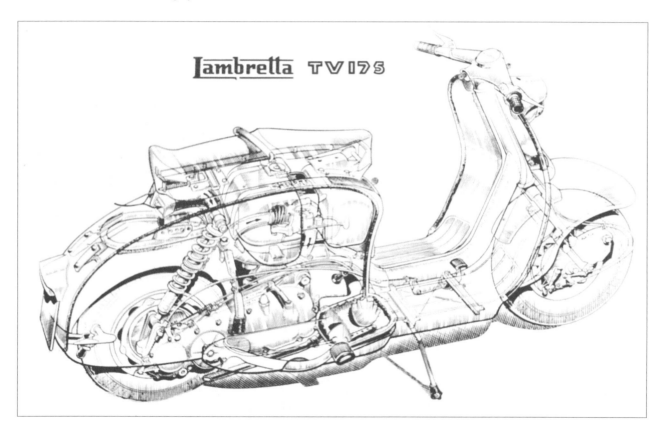

At first glance, the X-ray drawing of the TV 175 may seem similar to that of the Li version, but on closer inspection, it's clear to see the huge number of differences between the two models

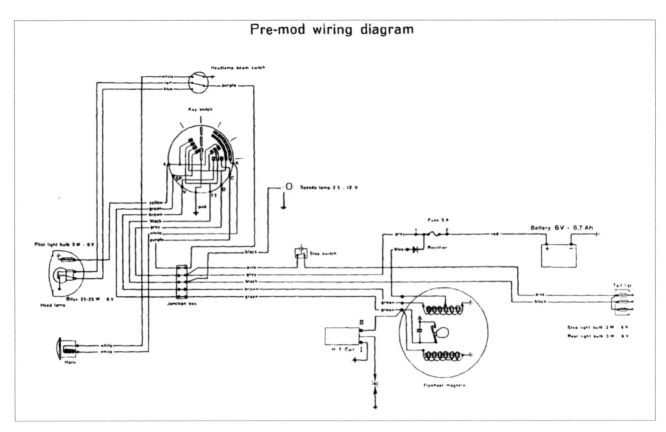

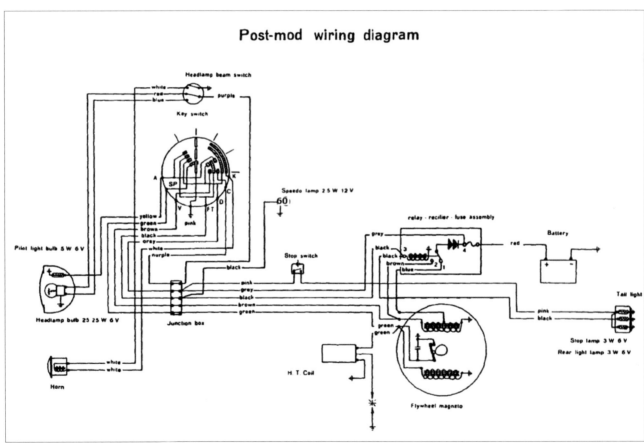

The wiring diagram differs from the Li range because the TV 175 is fitted with a key-operated control switch for the lights. The post-modification diagram has a relay fitted to the rectifier and fuse box

Lambretta Li 125-150 and TV175 technical details and specifications

DIMENSIONS

Wheelbase	1292 mm
Maximum length	1830 mm
Maximum width	710 mm
Maximum height	1060 mm
Carrying capacity	2

	Li 125	Li 150	TV 175
WEIGHT	104kg (230 lbs)	104kg (230 lbs)	120kg (264lbs)
PERFORMANCE (To CUNA rules)			
Maximum speed	48 mph/77 kph	54 mph/86 kph	64 mph/104 kph
Consumption	121mpg	120 mpg	94 mpg
Standing ¼ mile	N/A	N/A	N/A
ENGINE/TRANSMISSION GROUP	Li 125	Li 150	TV 175
Single-cylinder two-stroke			
Capacity	124cc	148 cc	170 cc
Bore	52 mm	57 mm	60 mm
Stroke	58 mm	58 mm	60 mm
Compression ratio	7.00:1	7.00:1	7.5:1
Maximum power (To CUNA rules)	5.20 BHP@5200 rpm	6.5 BHP@5300 rpm	8.60 BHP@6000 rpm

The cylinder is in cast iron with specially designed porting inclined horizontally. Cylinder head in light alloy pressure cast. Piston cast in light alloy. Connecting rod in high resistance steel, with brass bush small end. Crankshaft on ball and needle roller bearings. Ignition by flywheel magneto and external H.T coil. Fixed advanced ignition. Carburettor twist grip controlled. Cooling by forced air from flywheel fan. Aluminium induction manifold.

Clutch   Multiple discs in oil bath

Transmission   Duplex chain ratio 1-3.066 (15/46) between crankshaft and clutch sprocket (Li 125 and 150 only)

GEARBOX

4 speed, twist grip combined with clutch and controlled by double cable. Gear indicator on twist grip boss

Overall ratios	Li 125	Li 150	TV 175
	1st 17.40 – 1	1st 13.95 – 1	1st 14.32 – 1
	2nd 10.71 – 1	2nd 9.00 – 1	2nd 9.77 – 1
	3rd 7.47 – 1	3rd 6.67 - 1	3rd 7.30 --1
	4th 5.65– 1	4th 5.22 – 1	4th 5.69 – 1

LUBRICATION

Engine is lubricated by petrol/oil mixture in a 25 – 1, 4% ratio of two-stroke oils: transmission and gearbox by SAE30, the level of which is checked by the appropriate level plug

CARBURETTOR

Dellorto MB18 BS5 (Li125), MB19 BS5 (Li 150) MB23 BS5 (TV 175) Air filter diaphragm type; fitted with cable-operated starting device and large air intake at the rear of the frame

li125	li 150	TV 175
Choke – 18mm Main jet – 92	Choke – 19mm Main jet – 95	Choke – 23mm Main jet – 105
Slide – 50 Atomiser – 260 B	Slide – 50 Atomiser – 260 B	Slide – 70 Atomiser – 240 B

FRAME

Central beam in large section steel tubing welded to pressed steel rear section

SUSPENSION

Front - Trailing links and helical springs fitted in fork legs (with dampers on TV 175)

Rear - Swinging engine unit pivoting on silent block mountings controlled by combined helical spring and hydraulic shock-absorber unit

STEERING & WHEELS

Direct steering through fork stem in steel tubing. Interchangeable wheels with rims in pressed steel; 10" diameter. Anti-theft steering lock

Tyre diameter	3.50 x 10"	
Tyre Pressures	1 Person	2 Persons.
Front	18lb/sq"	18lb/sq"
Rear	28lb/sq"	34lb/sq"

BRAKES

Front   Expanding shoes, hand-controlled by cable drum 150 mm. diameter

Width of shoe lining - 25   Shoes cover two 120-degree sections

Rear   Expanding shoe pedal-controlled through cable drum 150 diameter

Width of shoe lining - 25 mm, Shoes cover two 120-degree sections

ELECTRICAL LAYOUT

Current supplied by 27w 4 pole magneto flywheel and Exide 3EV9.11 battery.

Round headlamp with dual filament bulb 6 v 25/25 w.

City light bulb 6 v 5W

Rear light complete with a red gem, bulb 6 v 3w x3 on TV 175.

Illuminated speedometer by 12 v 3w bulb.

DC horn

FUEL MIXTURE

Petroil - Standard petrol and SAE 40 or SAE 30 Two-stroke oil, first 900 miles 5%, After 900 miles 4%

TANK

Total fuel tank capacity of 8 litres (1.9 gallons), Reserve 1 litre (1.7 pints) TV 175 8.5 litres, reserve 1.2 litres

SILENCER

Double chamber expansion type with rock-wool soundproofing. Designed internally to give maximum power with low noise level. Maximum 86 dB.

TOOLS

Tool roll, containing:
- 1 – Spark plug spanner
- 1 – 8 x 10 mm spanner
- 1 – Screwdriver
- 1 – 10 mm Allen key
- 1 – 3.5 mm Allen key

Production numbers

LI 125.   1958: 500,000- 520,809

1959: 520,810- 547,747

Li 150.   1958: 500,000- 541,924

1959: 541,925- 610,944

TV 175.   1957: 001,001- 002,126

1958: 002,127- 011,088

Reprinted by kind permission of
www.britishlambrettaarchive.co.uk

Other publications in the Lambretta technical and history series are available direct on the Amazon platform worldwide

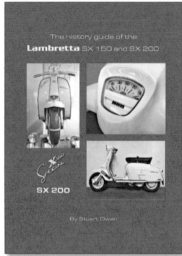

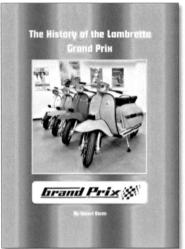
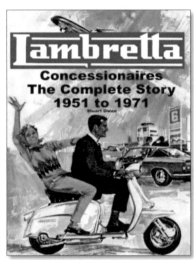
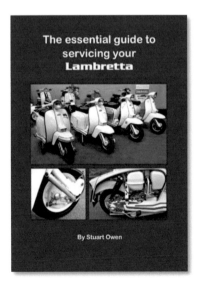
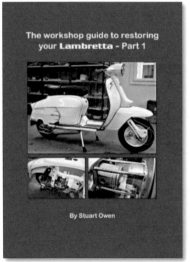
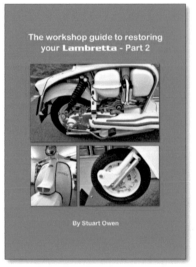

For more information on forthcoming titles in these series visit Facebook.com-The Lambretta history channel and Facebook.com-The Lambretta generation

Made in the USA
Columbia, SC
30 June 2025

60090525R00024